YOUR KNOWLEDGE HAS VALUE

- We will publish your bachelor's and
  master's thesis, essays and papers

- Your own eBook and book -
  sold worldwide in all relevant shops

- Earn money with each sale

Upload your text at www.GRIN.com
and publish for free

Bibliographic information published by the German National Library:

The German National Library lists this publication in the National Bibliography; detailed bibliographic data are available on the Internet at http://dnb.dnb.de .

Imprint:

Copyright © 2018 GRIN Verlag
Print and binding: Books on Demand GmbH, Norderstedt Germany
ISBN: 9783668702295

This book at GRIN:

https://www.grin.com/document/425564

Dr. Marshall Goldberg

The cellular automaton interpretation of aging and cancer

GRIN Verlag

The Cellular Automaton Interpretation of Aging and Cancer

2018

M. Goldberg

Table of contents

Introduction

Cells of multicellular organisms communicate with one another through gap junctions that control the movement of ions and other molecules from the cytoplasm of one cell to the cytoplasm of adjacent cells. Gap junctions are composed of connexin proteins [53]. We define a Gap Junction Bioelectric Network (GJBN) as the totality of gap junction communication among a mass of cells.

Cellular automata are mathematical models of many natural phenomena, including biologic processes such as shell patterns in mollusks [1]. Simple automaton rules can result in complexity, and there is no hierarchy of complexity once a threshold is reached [1, 3]. The threshold for complexity is low and rather easy to reach. More complex rules do not increase complexity. Wolfram cellular automaton #110 is the simplest such rule [1].

Cellular automaton #110 is complex, Turing complete, and capable of universal computation. It is bound by the Principle of Computational Equivalence (PCE), and the Principle of Computational Irreducibility (PCI) [1]. If the GJBN is a complex system, then the PCE holds that #110 is sufficiently complex to model it. The PCI teaches that there are no 'shortcut' formulas allowing one to calculate the future state of a complex automaton merely by plugging in a future time (Tfuture). Instead, one must 'run' the automaton to see its outcome. In the same way, GJBN modeled by cellular automaton #110 must be 'run' to determine the outcome of its computations.

This paper proposes that:

- Computations of the GJBN can be modeled by cellular automata.
- Entropic dysregulation of a complex GJBN results in aging and cancer.

'Cracking the bioelectric code' [7] implies there might be set of equations which can model the complex biochemical and bioelectric activities of a multicellular organism [5]. The PCI, however, implies there are no 'shortcut' equations able to adequately describe the complex biochemical interrelationships of an organism's environment, gap junctions, and genome. Instead, a class 4 Wolfram cellular automaton such as #110 is required and sufficient to model these relationships accurately and comprehensively [11, 13, 42, 44, 46, 58].

Symbolize a GJBN modeled by Wolfram cellular automaton #110, as **GJBN₁₁₀**. Gap junction conductance is regulated by the: **1.** production of connexins; **2.** numbers and types of connexin channels; **3.** ratio of connexin formation vs. destruction; **4.** percent of open connexin channels; **5.** voltage gating; **6.** dispersion and internalization of connexin channels into the cytoplasm; **7.** cell membrane potential (Vmem); **8.**phosphorylation; **9.**ion concentrations; **9.**antibodies [14, 15, 34, 43].

Cellular automaton #110 can emulate any other cellular automaton [1]. If by choosing a specific 'block' or 'neighborhood' of automaton cell values at time 'T', one can determine the subsequent value of a cell at time 'T+1', then #110 can emulate any another cellular automaton. Importantly, just as a 'block' of #110 cell values can be made to emulate any cellular automaton X, so too can $GJBN_{110}$ alter and combine its gap junction conductivities to emulate any $GJBN_X$.

Information flow in a cellular automaton depends on the Wolfram class of the automaton. For example, in cellular automaton #0, a class 1 cellular automaton, there is no information flow. In #30, a class 3 random cellular automaton, information flow is uncontrolled, and spreads to every part of the automaton. Information flow in #90 produces a nested pattern. In #110, a class 4 automaton, information flow is controlled and confined to moving and colliding structures.

It is important to note that $GJBN_{110}$ remains in overriding control even when it is emulating another cellular automaton.

Networks modeled by cellular automata are topologically robust. [8, 20, 27, 37, 64] These networks operate within a 'homeostatic range' with a well-defined mean, μ, and a narrow Vmem standard deviation, σ. See **Figure 3**.

The entire multicellular organism, modeled by #110, is symbolized as a scale-free gap junction bioelectric network by writing **GJBN*₁₁₀**, where the asterisk symbolizes all fractal scales of the organism: **1.** auto-catalyzing, self-organizing molecular networks which lie at the base of larger-scale structures such as DNA and its transcriptional availability as homochromatin; **2.** the mitochondrion and other cell organelles; **3.** the cell and its gap junctions, and; **4.** tissues, and organ systems. [8, 20, 29]

All of these scales are interlocked, each affecting and being affected by the others. They form a self-organizing, self-repairing, fractal, small-world network [8, 11, 23, 24, 27, 29, 30, 42, 45, 46, 65, 66].

Although communication among organ systems depends on hormones and other chemicals (e.g. acetylcholine), we still consider a multicellular organism as a Gap Junction Bioelectric Network because gap junctions ultimately connect the system at the cellular level. Let 'X' represent the Wolfram number of any 1-dimensional cellular automaton. Networks of biochemical reactions in GJBN*ₓ can be modeled as information flow gradients in cellular automaton X. Damage at any scale causes damage at all scales. See **Figure 1**.

Main Part

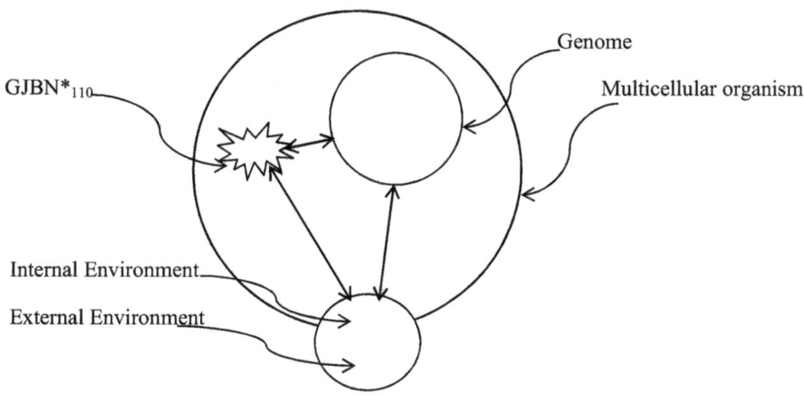

Figure 1

Figure 1 illustrates the scale-free, fractal structure of a multicellular organism. The complexity of the system requires class 4 cellular automaton #110 to model its behavior. The system is bound by the principle of computational irreducibility (PCI) meaning that 'shortcut' differential equations cannot comprehensively describe the behavior of the system.

Figure 2 illustrates how gap junction movements of negatively-charged chloride ions (electrons) determine cell membrane potential (Vmem), and Vmem gradients. Morphogens are signaling molecules that can activate various genes and gap junction conductivities depending on the morphogen's local concentration and its concentration gradient. Vmem gradients model morphogen gradients [16, 55].

Figure 3 illustrates Vmem control of cellular behavior [5]. Vmem affects the genome, and, reciprocally, the genome affects Vmem by governing the production and regulation of connexins, morphogens, and many other cell membrane components. Morphogens affect gap junction conductivities—allowing GJBN*110 to emulate any GJBN*x.

The basic rules governing GJBN*110 are simple just as cellular automaton rule #110 is simple, but the outcome of its computations are complex (PCE), and bound by the principle of computational irreducibility (PCI). This means that current models of embryogenesis, morphogenesis, aging, and cancer that are based on differential equations are at best only approximations, and cannot capture all the biochemical and computational complexities of a multicellular organism. Analysis of the complexities of a multicellular organism will require models based on GJBN*110. Inasmuch as simple rules can lead to complex results, development

5

of these computer models may not be as difficult as one might suppose. <u>A common mistake is making the model too complex</u> [11].

GJBN*$_{110}$ models the spatio-temporal patterns of Vmem potentials and Vmem gradients that define and guide the overall geometry of embryogenesis, morphogenesis, healing, regeneration, and the <u>dynamic equilibrium of morphostasis responsible for structural stability</u>. A youthful morphology depends on maintenance of morphostasis despite continual damage from entropy. Aging may be a consequence of loss of morphostasis (youthful structural stability) due to entropic degradation of GJBN*$_{110}$.

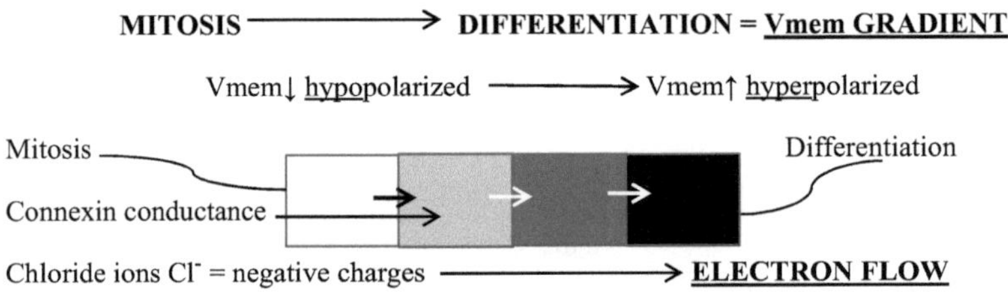

Figure 2

Figure 2 illustrates a row of biologic cells connected by gap junctions (short arrows) controlling the flow of chloride ions (e.g., electrons—negative charges) between the <u>cytoplasm</u> of cells. Gray-colored boxes represent cells with intermediate Vmem values depending on their cytoplasmic chloride ion concentration. If negatively-charged chloride ions move to the right, then cells losing cytoplasmic negative chloride ions exhibit <u>hypo</u>polarized membrane potentials with a lower Vmem and a tendency for mitosis, while cells receiving increased cytoplasmic negative charges become <u>hyper</u>polarized with a higher Vmem and a tendency for differentiation as illustrated in **Figure 3** below. <u>A Vmem gradient is established locally.</u> The slope of the gradient (shallow or steep) depends on gap junction conductivities. Because many types of ions can flow through gap junctions, then <u>a multitude of different Vmem gradients can be established</u>. The local concentration of a morphogen activates specific homeobox genes. <u>In a 3-dimensional syncytium of biological cells, a morphogen gradient may not be identical in every direction—the system is anisotropic.</u> Thus, body structures can have different cross-sectional shapes along each dimension.

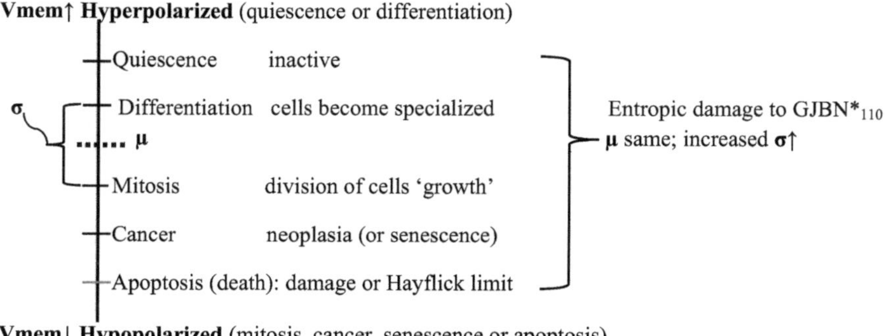

Figure 3

Figure 3 illustrates the relationship of Vmem and cellular behavior [5]. If a cell is maximally hyperpolarized, it becomes quiescent. If Vmem is less hyperpolarized, the cell might differentiate into an adult cell. At lower Vmem values, mitosis occurs. At an even more hypopolarized Vmem, the cell might become neoplastic or senescent, until at the lowest levels of hypopolarization, the cell can die. Low Vmem values favor cancer, unbridled mitosis, apoptosis, and perhaps senescent cells (neoplasia protection?). Vmem values probably determine which morphogens are produced by the cell; conversely, morphogens can affect Vmem values.

Spatio-temporal Vmem gradients and morphogen gradients of biologic cells can be modeled by GJBN*110. In **Figure 3** above, the dotted line within the smaller bracket indicates the 'normal' homeostatic mean Vmem value, μ, of cells. The smaller bracket indicates the 'normal' homeostatic range of the standard deviation, σ, of cells when they are controlled by GJBN*110. If the homeostasis of GJBN*110 is dysregulated by entropic damage, then the standard deviation σ of Vmem increases, σ↑ (larger bracket). Cells undergoing excessive, unnecessary, and uncontrolled mitoses may reach the Hayflick limit prematurely and die. Organisms appear to be trapped between cancer and senescence—a trade-off between neoplasia and aging. With loss of GJBN*110 control, and an increase in the standard deviation (σ↑), low Vmem cells can be at risk for cancer. Cellular senescence is a mechanism that may help prevent cancer. If entropic damage results in cells becoming more independent (unicellular), and less under the control of a gap junction bioelectric network, then there is an increase in the standard deviation (σ↑) of Vmem.

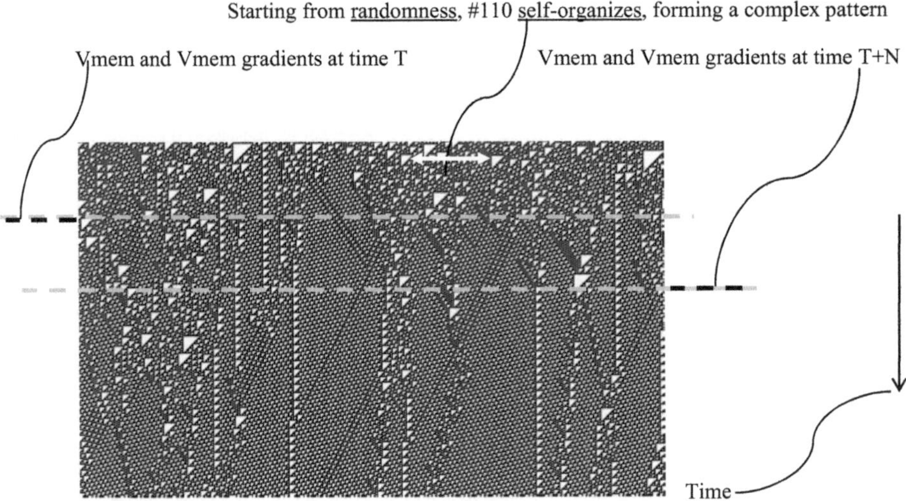

Figure 4

Figure 4 illustrates a portion of Wolfram complex cellular automaton #110 [1]. A horizontal row of the automaton models the spatial distribution of scale-free Vmem values and Vmem gradients of biologic cells at time T. Reading <u>vertically</u> down the cellular automaton illustrates the temporal changes in <u>scale-free</u> Vmem values and gradients from T to T+N. Control of connexins and gap junction conductance (e.g. by morphogens) allows $GJBN*_{110}$ to emulate any $GJBN*_X$. These <u>emulations are dynamic,</u> occur in 'real time,' and are epigenetic. Over longer time scales they interact with the genome and the environment, and emulations change as the organism develops, grows, heals, and ages. Morphogen concentrations and gradients can be modeled by $GJBN*_{110}$ or one of its emulations [46].

Figure 4 illustrates how, starting with an initial random pattern, cellular automaton #110 quickly self-organizes into a complex pattern [8, 20, 48, 64, 65, 66]. <u>Note that if one were to 'insert' a new row with a random pattern into any row of #110, the automaton would once again quickly reorganize itself to form a complex pattern.</u> Consequently, if there is entropic damage to $GJBN*_{110}$, it can also self-repair and reorganize itself [11, 20, 27, 37]. See **Figures 5a and 5b below.**

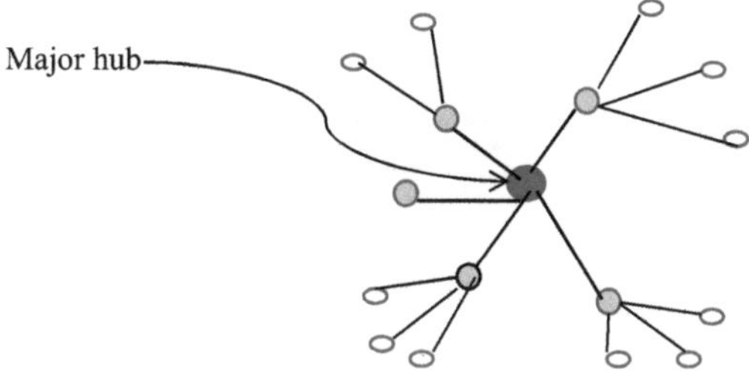

Figure 5a

Figure 5a shows a portion of a small-world network. The black node is a major hub.

Figure 5b

In **Figure 5b** the major hub and its connections illustrated in 5a have been lost as depicted by the lighter colors. The network has repaired itself by forming new connections that 'bridge' the gap created by the loss. The 'bridges' indicate repair of the network with a <u>virtual node</u> [6, 8, 11, 13, 20, 23, 24, 25, 27, 29, 31, 32, 42] .

In a cellular automaton, let a darker cell color (Figure 2 above) indicate a <u>hyperpolarized</u> Vmem leading to differentiation, and a lighter cell color indicate a <u>hypopolarized</u> Vmem leading to mitosis. Increasing the number of cell colors (as in a **totalistic cellular automaton** [1]) produces no new cellular automaton patterns so that if biologic cells have multiple Vmem values, then GJBN*$_{110}$ can still serve as a model.

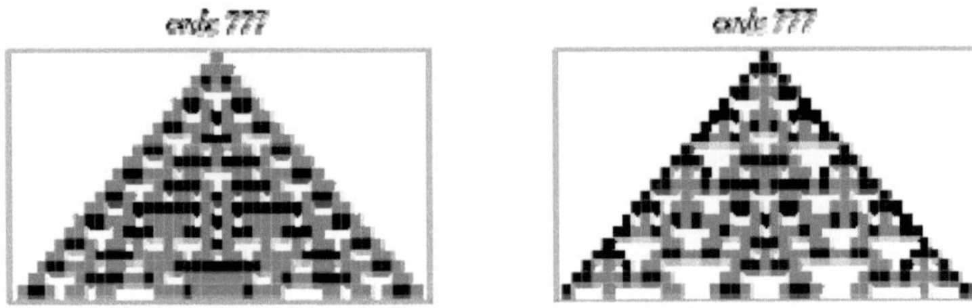

Figure 6

Figure 6 illustrates a <u>totalistic cellular automaton</u> [1] in which the color of the cell is the 'average' of the colors of 'neighborhood' cells. Note that these cells of varying colors can represent the various Vmem values of biologic cells in a gap junction bioelectric network, as in Figure 2 above. Therefore, a one-dimensional, two-color cellular automaton can model a 3-dimensional, multi-color cellular automaton. <u>This means that a 3-dimensional mass of biologic cells with many Vmem values and many anisotropic morphogen gradients can be modeled by a 1-dimensional 2-color cellular automaton and its emulations.</u> The basic patterns in 1-dimensional cellular automata are representative of all patterns seen in higher dimensions.

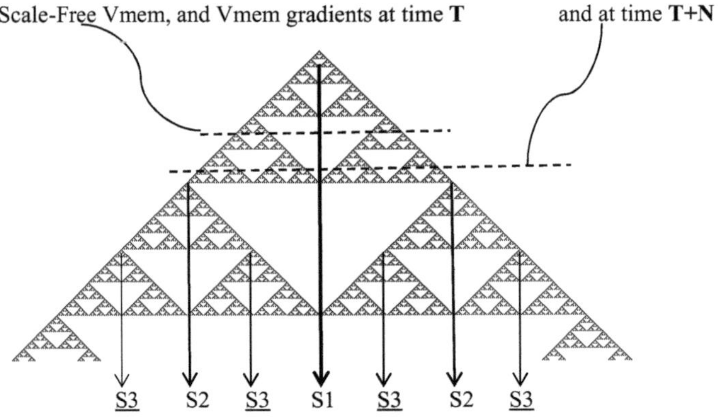

Figure 7

Figure 7 illustrates the nested, fractal pattern of cellular automaton #90 (Sierpinski triangle) with self-similarity at all scales. Cellular automaton #110 can <u>emulate</u> cellular automaton #90 by combining several cells such that a 'block' (neighborhood) of automaton cells determines the value of a subsequent automaton cell. Similarly, <u>GJBN*$_{110}$ can emulate GJBN*$_{90}$</u>, by altering gap junction connectivity such that a 'block' of biologic cells determines the subsequent Vmem value of a biologic cell [77]. <u>If morphogens can alter the conductivities of gap junctions, then it</u>

follows that as a morphogen diffuses through a mass of cells it could allow GJBN*$_{110}$ to emulate GJBN*$_{90}$. Increasing gap junction conductivities could result in a biological cell having a Vmem that is some function of the biological cells in its neighborhood—a criterion for emulation similar to that which occurs in a cellular automaton. The vertical arrows in Figure 7 show growth of light and dark regions as time passes. In this two-color 1-dimensional cellular automaton, white represents mitosis and black represents differentiation (see Figure 3). Depending on the 'scale' or nesting of the black (differentiating) cellular automaton cell (and local morphogen gradient), a nested gene from the same scale is 'called' or activated. This means that genes for different tissues such as bone, muscle, etc., as well as morphogens are activated at the correct location, scale, and time. Thus, this cellular automaton model can also account for the scale invariance of growth. Genes are nested or scaled in the sense that different parts of the genome 'program' are called or activated depending on the scale (as in Figure 7, relative to rest of organism) of the structure being formed [74, 75, 76]. Also note in Figure 7 that black and white cellular automaton cells are mixed at all scales so that mitosis and differentiation also occur together at all scales. The scale-free (scale invariant) nature of the system means that a simple diffusion model of morphogen gradients will have to be modeled as a scaled system [75, 76]. Observe how the system in Figure 7 doubles the structures at each iteration (for S1 there are two S2; for S2 there are four S3 structures—Figure 8 below) [76]. Thus, GJBN*$_{90}$ can model the scale-free, fractal 'tree' structure seen in many biologic structures such as limbs, and lung and vascular systems illustrated in **Figure 8**.

Figure 8

Figure 8 illustrates GJBN*$_{110}$ emulating GJBN*$_{90}$, producing a 'nested' Vmem gradient field, modeling a 3-dimensional, nested, anisotropic morphogen gradient field, which determines the nested 3-dimensional structure of limbs, and other 'tree' structures such as lungs, and vascular systems. Moreover, lower Vmem values with increased mitotic activity in a group of cells results in differential growth rates, producing many variations in the 3-dimensional shapes, and forms of various structures (folding, invagination, etc.). In Figure 7 above, there is a morphogen gradient along S1, another gradient along S2, and yet another along S3. This might mean that the original gradient along S1 is possibly picked up, modified, amplified, and renewed to form the gradient along S2. The S2 gradient is treated similarly to form gradient S3. Thus, morphogen gradients are possibly modified, amplified, and 'renewed' at each scale, thereby yielding primary, secondary, and tertiary morphogen species. This, rather than a single morphogen gradient, may allow for growth and specific tissue-type differentiation at each location and scale of the organism. In addition, a multicellular organism must interact with internal and external forces. Therefore, development and modeling of bone and blood vessels is guided by mechanical strain and hydrodynamic forces, respectively. Mechanical strain and hydrodynamic forces can affect Vmem and gap junctions because distortion from external or internal (growth) forces can alter cell membrane permeability.

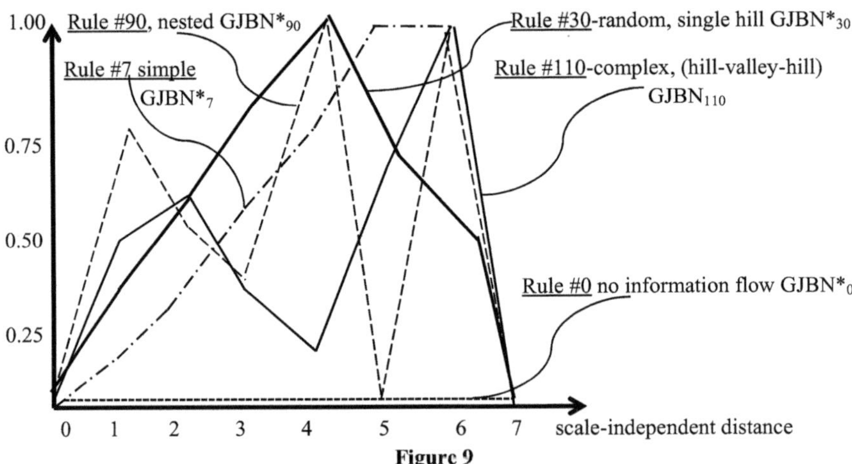

Figure 9

Figure 9 illustrates the asymmetry-graphs (A-graphs) of #110, #90, #30, #7, and #0[2]. Between any two scale-independent distances (1-2, 2-3, 4-5, etc.) the <u>slope</u> of the A-graph defines a Vmem gradient (positive slope = gradient in one direction; negative slope = gradient in the opposite direction). <u>A Vmem gradient can model a morphogen gradient</u>. In a cellular automaton the distance between any two points is <u>scale-free</u> so that a change at one scale is also a <u>change at different scales and, therefore, a change over distance</u>. The A-graph [2, 72] is a scale-free map of spatio-temporal Vmem gradients and morphogen gradients in a 3-dimensional biological cell syncytium. Morphogen gradients and local concentrations of morphogens control genome activation. <u>Morphogen gradients have to be scale invariant to allow for an organism's growth</u>. The genomic production of various types of connexins forming GJBN*110, can control intercellular gap junction communication among cells of the network over longer time periods. However, over shorter periods of time, GJBN*110 is a quasi-independent epigenetic control system. In Figure 9 above many other A-graphs not shown have <u>intermediate shapes</u> representing many Vmem gradients and different morphogen gradient patterns and structures. All of these intermediate A-graph shapes can be emulated by #110. They represent many different combinations of order, randomness, and complexity. Because many morphogens are cell products, it follows that morphogen gradients do not necessarily require a single 'center' for morphogen production in order for morphogen gradients to form. Rather, morphogen concentrations and gradients are probably scale invariant [76] and likely determined by local

Vmem values and Vmem gradients modeled by a cellular automaton. **Figure 2** above, for example, illustrates how a Vmem gradient can be established by <u>local</u> gap junction ion (<u>electron</u>) flow. Embryogenesis, morphogenesis, and morphostasis depend on scale-free, spatio-temporal morphogen gradients and the local concentrations of many primary and secondary morphogens [55]. The shapes of the A-graphs in Figure 9, above, and their <u>slopes</u> represent scale-free Vmem gradients that can model <u>scale-free morphogen gradients</u>. The ability of cellular automaton #110 and GJBN*$_{110}$ to <u>emulate</u> all of these A-graphs means that GJBN*$_{110}$ can emulate and model a multitude of Vmem gradients and morphogen gradients needed to determine and control the cellular activity and overall <u>geometry</u> of a multicellular organism. <u>Aging and cancer are most likely disorders involving disruption of an organism's geometry due to entropic dysregulation of</u> <u>GJBN*$_{110}$</u> [9, 18, 19, 56]. Note that A-graph #30 in Figure 9 above approximates a discrete version of a normal distribution curve. **Figure 10** below illustrates that Vmem and morphogen gradients in GJBN*$_{30}$ (as modeled by cellular automaton and A-graph #30) are fairly <u>symmetrically</u> distributed about a mean μ at all scales; and, therefore, tend to <u>cancel out</u> so that the geometry of a multicellular organism cannot be modeled or maintained. By contrast, other A-graphs have <u>asymmetric</u> Vmem gradients that <u>can</u> model various morphogen gradient geometries of a multicellular organism. Entropic <u>dysregulation</u> of GJBN*$_{110}$ to GJBN*$_{30}$, and the entropy theory of aging suggest an analogy with the <u>central limit theorem</u> and the second law of thermodynamics, applied in this case to a scale-free, small-world network GJBN*$_{110}$. It is proposed that aging and cancer are due to chronic <u>entropic dysregulation of a **network** of cells</u> <u>rather than damage to single cells</u>. **Figure 10** below illustrates the relationship of the entropic dysregulation of GJBN*$_{110}$ and the Central Limit Theorem.

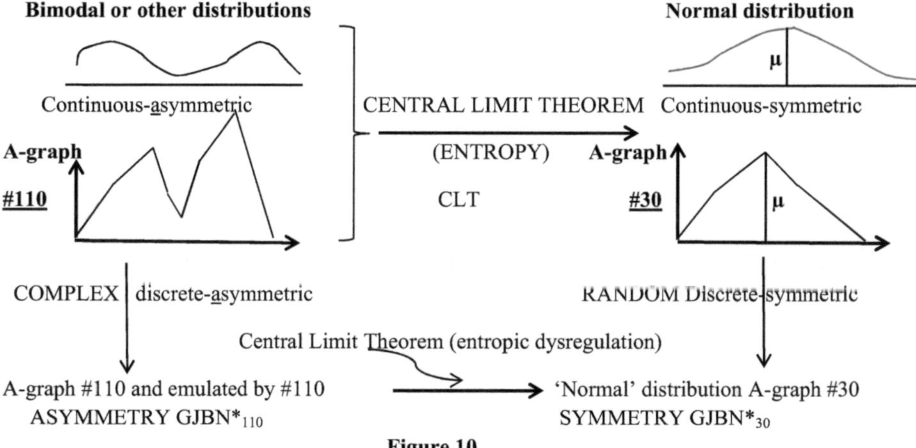

Figure 10

Figure 10 illustrates the putative relationship of the central limit theorem to <u>entropic</u> <u>dysregulation</u> of the complex GJBN*$_{110}$ into the random GJBN*$_{30}$. The <u>bimodal</u> A-graph of #110 is a discrete version of a continuous bimodal distribution. The unimodal A-graph of #30 is a discrete version of a continuous unimodal 'bell' curve. In probability theory, the central limit theorem (CLT) states that when independent random variables are added, their properly normalized sum tends toward a bell curve, even if the original variables are not distributed normally. This theorem also applies to discrete distributions (bar graphs, for example). Similarly, when the complex bimodal #110 A-graph is acted on by entropy (entropic dysregulation), random unimodal #30 A-graph is the result. The central limit theorem relates entropy, information, and the second law of thermodynamics. Entropy causes dysregulation of the asymmetry of the complex GJBN*$_{110}$ resulting in the symmetry of the random GJBN*$_{30}$. Thus, entropic dysregulation results in loss of the asymmetry of Vmem values and gradients, and, therefore, the asymmetric morphogen concentrations and gradients required for a multicellular organism's morphogenesis and morphostasis (morphostabiliy—stable geometry). Symmetry leads to randomness. Aging and cancer are sequelae of dysregulation of a multicellular organism's geometric stability [7].

In GJBN*$_{110}$ as in A-graph #110 we have [3, 4, 21]:
Asymmetry + Randomization (entropy-computation) → Complexity

In GJBN*$_{30}$, as in A-graph #30 we have:
Symmetry + Randomization (entropy-computation) → Randomness

The emergence of order or complexity [4] in this situation prompted Schrödinger to propose the concept of negative entropy or negentropy, the ability of a living organism to export entropy out of its system in order to keep its entropy low and maintain order [48]. Complexity may arise from entropy acting on asymmetric substrates [4]. Complexity is an emergent property arising from scale-free partitioning of randomness [2]. If 'entropy' represents any possible damage (physical, bacterial, viral, chemical, radiation, anoxic) at any scale to a multicellular organism then,

Entropic dysregulation and randomization

GJBN*$_{110}$ ⟶ GJBN*$_{30}$

Aging and death are due to gradual entropic dysregulation of complex GJBN*$_{110}$ until it becomes a random network modeled by GJBN*$_{30}$.

Aging may be related to the accumulation of senescent cells and the spread of damage from these cells to healthy cells. Senescent cells have a 'senescent-associated secretory phenotype' (SASP) [54, 59]. They are metabolically active, and can damage and poison other cells with cytokine-induced-inflammation. Relatively few senescent cells can produce a lot of damage. Although senescent cell cytokines can probably diffuse over shorter distances, transmission of damaging effects over longer distances probably requires intact gap junctions [19, 36, 49, 50, 52, 56].

If GJBN*$_{110}$ is able to emulate GJBN*$_0$—*recall there is no information flow in Wolfram cellular automaton #0*—then senescent and cancer ('instructor' cell [71]) 'damage' information transmission over longer distances is blocked, thereby avoiding the 'bystander effect' [49, 50, 57]. Because GJBN*$_{30}$ cannot emulate GJBN*$_0$ and because cellular automaton #30 allows uncontrolled transmission of information, then senescent and cancer cell 'damage' information can flow unimpeded to viable cells [56, 57].

In captivity, naked mole rats can live more than 30 years [51]. Most rodents live only three years. Aging in the naked mole rat is apparently delayed by controlling the metabolism of senescent cells—senescent cell genes in the mole rat are more organized than in short-lived rodents.

Consequently, control or prevention of aging and cancer might require: **1.** Maintaining GJBN*$_{110}$ so that it can emulate GJBN*$_{90}$ and, thereby, prevent the formation of cancer cells. **2.** Emulating GJBN*$_0$ and halting the spread of 'damage information' from senescent and cancer ('instructor') cells to undamaged cells [69]; **3.** Removal of senescent cells before they can damage stem and pluripotential cells [49-52].

Aging has been considered to be the result of accumulated entropic damage to an organism, but how is aging related to entropy? In <u>unicellular</u> organisms, damage to a cell often results in its death. For example, we expect an effective antibiotic to kill every bacterial cell it contacts. For single cells, entropy and death are directly causally related. <u>Unicellular organisms die, but do not age.</u>

In engineering one strives to design perfect parts, and then combine them into a machine with minimal global imperfections. Nature, however, seems to take a different and perhaps simpler approach, namely constructing global 'perfection' by utilizing a multitude of redundant but less than perfect parts. Accordingly, there may be a vast <u>redundancy</u> of imperfect parts in multicellular organisms. Redundant imperfection might add up to global perfection if imperfect 'parts' are <u>randomly</u> distributed about a 'center of perfection' such that these randomly-distributed imperfections cancel each other globally. Microscopic randomness tends to 'wash out' allowing macroscopic fidelity in morphogenesis.

Gradual loss of redundancy may correlate with aging in multicellular organisms [38]. From the perspective of the ideas in this paper, aging is related to 'loss of redundancy' because of persistent entropic damage to GJBN*$_{110}$. Multicellular organisms trade individual cell autonomy for group bio-stability. Single ants, bees, termites, wolves, and humans, are more viable in the plurality rather than as non-cooperating individuals. Imperfections in individuals are 'corrected' or off-set by random opposing imperfections in others so that as a group, individual random imperfections are minimized. The standard deviation of the group is less than that of the individual. However, creativity in humans must not be thwarted. Note that while GJBN*$_0$ can block the flow of damage information to healthy cells it is not the same as having <u>no</u> gap junctions, i.e. a collection of individual cells acting independently. The importance of this distinction is that total loss of gap junction control of cells renders these cells independent, i.e., like unicellular organisms. Cancer cells are no longer constrained to be part of a network; they are freed from network control (loss of 'contact inhibition'). As unicellular organisms, cancer cells survive at the expense of the 'host'.

In multicellular organisms, entropic damage to a cell or even groups of cells is probably not the proximate cause of the changes in morphology known as aging. The <u>Hayflick</u> limit, telomeres, and telomerase are probably <u>not</u> primary factors in aging. Instead, the Hayflick limit determines the maximum number of mitoses an <u>unbridled</u> cell may undergo when it is released from network control, and the group homeostasis, regulation, and 'safety' of GJBN*$_{110}$. <u>An intact GJBN*$_{110}$ protects cells from neoplasia, and unnecessary mitosis</u>. For example, it has been shown that loss of network control by chemical inhibition of connexins Cx43 and Cx32 can initiate cancer [70]. Moreover, scars and keloids have decreased gap junction activity and an increased risk for both primary cancer and as sites for metastatic cancer [73]. Chronically inflamed or damaged tissues may also have similar risks.

At larger inter-organ scales, other links can be lost: nervous system connections from stroke, endocrine disease, etc. Nevertheless, entropic damage to the network at any scale ultimately involves gap junctions. Following entropic damage, the remaining undamaged portion of the network reorganizes to form a new GJBN*$_{110}$ in several steps:

- There is direct quarantine of information associated with the damaged subset of the network because the subset's gap junction links to the undamaged portion of the system have been disrupted.
- 'Bystander damage' to the rest of the system is prevented if GJBN*$_{110}$ emulates GJBN*$_0$, and blocks gap junction transmission of 'damage' information [52].
- Damaged cells are removed.
- Pluripotential cells are replaced.
- The GJBN*$_{110}$ reorganizes itself into a 'new' GJBN*$_{110}$ network [20, 27, 37].

A systems theory of aging [38] suggests that 'age' is not simply a 'wearing out' in the same sense that an inanimate object wears out; instead, 'aging' is embodied in the common expression, '<u>growing</u> old.' Thus, when an organism's growth is complete, there a gradual entropic change from a complex to a random biologic network resulting in a gradual loss of the asymmetric morphogen gradients necessary for maintenance of a youthful morphology. *The difference in appearance between a young and aged person may be due to a <u>gradual</u> entropy-driven change in network topology from a complex GJBN*$_{110}$ to a random GJBN*$_{30}$—the organism 'grows' old.*

Discussion

When unicellular organisms became multicellular, they had to give up their individual behaviors in support of a coordinated group of cells—a gap junction bioelectric network. Individual cell death was replaced by aging and cancer. <u>A gap junction bioelectric network can be seen as a biological device able to control anisotropic, scale-invariant electron (ion) flow in three dimensions</u>. The Principle of Computational Equivalence (PCE) teaches that complexity is easy to obtain. It arises from simple rules, and once a rather low threshold for complexity is reached there is no hierarchy of complexity—all complexities are equivalent. Though the rules governing the gap junction bioelectric network may be simple, a complex system is required to describe its behavior. Wolfram cellular automaton #110 can model the gap junction bioelectric network.

The Principle of Computational Irreducibility (PCI), a corollary of the PCE, holds that a 'shortcut' differential equation representation of the $GJBN*_{110}$ is inherently incomplete [1,22]. Instead one must use $GJBN*_{110}$, and watch it 'compute' to observe its outcome. Complex systems are deterministic, yet *a priori* indeterminate. A common mistake in setting up models is that they are made unnecessarily complicated. Often the system being modeled has behaviors not covered by the original model so that 'corrections' or additions have to be made. Use of the $GJBH*_{110}$ at the outset can avoid these shortcomings since it can have all the necessary complexity to account for embryogenesis, morphogenesis, growth, healing, cancer, and aging.

This paper proposes that aging and cancer are due to <u>entropic dysregulation</u> of the gap junction bioelectric network. Entropic dysregulation of a complex $GJBN*_{110}$ causes it to become a random $GJBN*_{30}$. Entropic damage at any scale affects all scales. Aging and cancer involve properties of the <u>entire</u> scale-free network. This means that cancer cannot be fully understood in terms of damage to individual cells or DNA. Cancer and aging have to be analyzed from a systems approach as dysregulation at all scales of a complex, dynamic, small-world network. Vmem and Vmem gradients are epigenetic; the underlying genome is free to evolve independently.   Entropic damage to an organism causes gradual dysregulation of $GJBN*_{110}$ and its ability to emulate other automata, such as $GJBN*_{90}$. Loss of morphostasis (geometric stability), rather than the Hayflick limit, might be the proximate cause of aging. If entropic dysregulation of the $GJBN*_{110}$ yields $GJBN*_{30}$, then complex Vmem gradients and morphogen gradients become <u>randomized</u>. Geometric stability is lost, and aging and cancer occur. *The difference in appearance between a young and aged person may be due to a <u>gradual</u> change in network topology from a complex $GJBN*_{110}$ to a random $GJBN*_{30}$.* Entropic damage leads to quarantines and repairs which can be visualized as node loss, rewiring, and reorganization to form a new $GJBN*_{110}$ topology. Aging occurs when the system loses its capacity to quarantine damage, 'rewire,' and reorganize $GJBN*_{110}$ [20,27,28, 66]. If the network becomes random, <u>symmetric</u>, and incapable of universal computation and emulation, then it cannot support morphostasis (morphostability) or prevent primary and metastatic cancer. Morphostasis is the ability of GJBN*$_{110}$ to emulate GJBN*$_{90}$ and other automata in order to maintain the <u>dynamic equilibrium of youthful structures</u>.

A cellular automaton is <u>not</u> a simple 'picture' of the <u>morphology</u> of a multicellular organism. Instead, it is a model of the spatio-temporal Vmem gradients and three-dimensional anisotropic morphogen gradients of a multicellular organism. Morphogen gradient anisotropy is required because 3-dimensional structures can have different cross-sectional shapes along each dimension. Additionally, morphogen gradients must be scale free in order to model the morphology of an organism over a range of scales as its size increases during growth. <u>The scale-free structure of cellular automata can model the scale-free invariance of morphogen gradients</u>. Morphogen gradients have to be both anisotropic and scale invariant in order to model the development of growing, 3-dimensional structures. Accordingly, morphogen and gradient diffusion models have to include anisotropic gap junction transport of morphogens and other variations of morphogen transport including local production, renewal, amplification, variation, and inhibition. Current models involving simple Turing diffusion of morphogens have to be modified to allow for anisotropy and scale invariance.

The global behavior of a scale-free cellular automaton is under the control of a <u>local rule</u> so that any changes to the automaton at one scale at time T will be reflected at <u>all scales</u> in the next time cycle of the automaton at time T+1. Therefore, a change at a small scale will also be seen at larger scales [6]. An observer, unaware of the fractal nature of the cellular automaton, might infer that the change had been transmitted instantaneously to a 'distant' part of the automaton. A change<u> of scale might be confused with a change of distance</u>, leading to a paradox of 'entanglement.' If the universe can be described as computations of a cellular automaton, then 'entanglement' might be due to confusing a change of scale with a change in distance [1,41,60,61].

If Nature's organizing principle is something like a cellular automaton, then 'a law of nature' might be described as a scale-independent <u>computational relationship</u>. A 'law' or computational relationship discovered at one scale will be similar in some way at all scales—*Fractal Principle*. Should we replace the notion of 'natural laws' embodied in equations with the concept of the <u>scale invariance of computational relationships</u>?

Will he laws of nature have to be reformulated so that they follow the fractal principle, and are scale invariant? Quantum mechanics seems to avoid this principle unless nature represents the computations of some kind of cellular automaton. Wolfram, Wheeler ('it from bit'), and 'tHooft have suggested that one can think of the universe and all its manifestations, including life, as computations of some kind of cellular automaton. At the quantum level, there may be no-thing—nothing but computation [41,60, 61]. Could quantum mechanics turn out to be deterministic and related to superdeterminism and counterfactual definiteness as Gerard 'tHooft has suggested [41,61]?

Entropic dysregulation of network homeostasis with an increase in the standard deviation ($\sigma\uparrow$) of Vmem can result in cancer, cell senescence, and Hayflick limit apoptosis. An increase in standard deviation due to entropy is akin to loss of network control so that cells behave (and 'misbehave') as if they were independent unicellular organisms. In this regard, cellular senescence is a protective mechanism.

Several interesting ideas and questions are suggested:

- Aging and cancer might be prevented or reversed by maintaining or restoring $GJBN^*_{110}$. Electroceuticals are molecules that might be used to alter Vmem. Can $GJBN^*_{110}$ be maintained or restored with electroceuticals? Current ideas about anti-aging therapies involve telomeres and telomerase. However, system-based computational models that seek to maintain and restore $GJBN^*_{110}$ could offer more effective anti-aging treatments.

- Stem cell regeneration is an area of interest. There is evidence that adult somatic cells can be revamped as multi-potential cells [63]. Are stem cell maintenance, restoration, and recruitment controlled by Vmem and Vmem gradients and dependent upon an intact $GJBN^*_{110}$?

- Can the 'biologic age' of a tissue be determined using voltage-sensitive dyes to measure the real-time spatio-temporal picture of the reaction patterns of Vmem gradients in response to electrical stimulation of one region of the tissue? A dynamic spatio-temporal reaction pattern of Vmem changes might indicate whether a particular tissue is governed by cellular automata #110, #90, #30, #0 or whether there is no network at all. Absence of $GJBN^*_{110}$, and the presence of $GJBN^*_{30}$ or even no network connectivity might be a measure of the aging and neoplastic liability of the tissue.

- Neoplastic cells have hypopolarized Vmem values. Could the use of Vmem-sensitive fluorescent dyes in situ or in surgical biopsies provide a technique for quick intraoperative evaluation of fresh tissue biopsy margins for cancer?

- Could a cellular automaton model of a virus be able to compute subtle and unexpected structural changes in the virus, so that vaccines might be produced in advance of an epidemic?

- Is our inability to foresee or 'pre-calculate' negative outcomes in war, business, law, and government regulations ('the law of unintended consequences'), an effect of the PCE and the PCI? Perhaps cellular automaton models based on Wolfram #110 might allow us to pre-calculate and avoid these unexpected results.

Computation, as an abstraction, is a general phenomenon. It can be embodied in cellular automata and the networks they model. 'Computation' is independent of the particular entities that are involved [1]. 'Life' is best understood and analyzed as computation, rather than biochemistry.

If in living organisms, information and computation are primary while the specific chemicals involved are secondary, then life might evolve from many different chemistries in vastly diverse environments. Self-organizing, auto-catalyzing, information-processing small-world networks could be the key to the origin and maintenance of life [64-67].

Because of the PCE, complexity, life, intelligence, and consciousness throughout the universe may not only be common, but inevitable. Does the PCI prevent us from predicting the future course of evolution or of the universe [62]? Is there a limit to human knowledge?

Summary

Cells of multicellular organisms communicate with one another through gap junctions that control the movement of ions and other molecules from the cytoplasm of one cell to the cytoplasm of adjacent cells. Gap junctions are composed of connexin proteins [53]. A Gap Junction Bioelectric Network (GJBN) symbolizes the totality of gap junction communication among a mass of cells. Computations of the GJBN can be modeled by cellular automata. The entire multicellular organism, modeled by complex Wolfram cellular automaton #110 is a scale-free gap junction bioelectric network symbolized as GJBN*$_{110}$. Simple rules and models based on complex cellular automata can model complex biochemical networks in multicellular organisms. Cellular automaton #110 can emulate any other cellular automaton [1]. A 1-dimensional cellular automaton can model a 3-dimensional cellular automaton. In a 3-dimensional syncytium of biological cells, a morphogen gradient may not be identical in every direction—the system is anisotropic. The PCI teaches that there are no 'shortcut' formulas allowing one to calculate the future state of a complex automaton merely by plugging in a future time (Tfuture). Instead, one must 'run' the automaton to see its outcome. In the same way, GJBN*$_{110}$ must be 'run' to determine the outcome of its computations. The system is deterministic, yet *a priori* indeterminate. GJBN*$_{110}$ can emulate GJBN*$_{90}$, by altering its gap junction conductances such that a 'block' or neighborhood of biologic cells determines the subsequent Vmem value of a biologic cell. For example, GJBN*$_{110}$ can emulate GJBN*$_{90}$, thus producing a 'nested' Vmem gradient field. This nested field can model a 3-dimensional, nested, anisotropic, scale-invariant morphogen gradient field that determines the nested 3-dimensional structure of limbs, and other 'tree' structures such as lungs, and vascular systems. Anisotropy is required since structures in a multicellular organism have different cross sectional shapes along each dimension. The GJBN*$_{110}$ system is scale-free. Accordingly, morphogen gradients must be scale free or scale invariant to allow for the growth of an organism. Simple morphogen diffusion models have to be modified to account for scale invariance. Morphogen concentrations and gradients can be modeled by GJBN*$_{110}$ or one of its emulations [46]. A Vmem gradient can be established by local gap junction ion (electron) flow. Low Vmem values favor cancer, unbridled mitosis, apoptosis, and perhaps senescent cells (neoplasia protection?). An increase in the standard deviation ($\sigma\uparrow$) of Vmem occurs if entropic damage results in cells becoming more independent, and less under the control of a gap junction network. Networks of biochemical reactions can be modeled as information flow gradients in a cellular automaton. Damage at any scale causes damage to the network at all scales. The PCI indicates that current models of embryogenesis, morphogenesis, aging, and cancer that are based on differential equations are, at best, only approximations, and cannot capture all the biochemical and computational complexities of a multicellular organism. Entropic dysregulation of GJBN*$_{110}$ results in aging and cancer. *Aging and cancer are both results of the dysregulation of the stability of the geometric form of an organism* [7]. The Hayflick limit, telomeres, and telomerase are probably not the primary factors in aging. It is suggested that loss of morphostasis (geometric stability), rather than the Hayflick limit is the proximate cause of aging. Aging is related to 'loss of redundancy' because of persistent entropic damage to the GJBN*$_{110}$ system. Cancer and aging have to be analyzed from a systems approach as <u>dysregulation at all scales of a complex, dynamic, small-world network</u>. The entropy theory of aging suggests an analogy with the <u>central limit theorem</u> and the second law of thermodynamics applied in this case to a scale-free, small-world <u>network</u>, GJBN*$_{110}$. Simple rules can lead to complexity. If living systems are best described in terms of computation rather than biochemistry, then life might evolve inexorably from many chemistries in vastly diverse and non-Earth-like environments.

References

1. Wolfram, S., *A New Kind of Science* ©2002 Wolfram Media Inc. Wolfram, Stephen, Wolfram Media, Inc., May 14, 2002, ISBN 1-57955-008-8.

2. Goldberg, M. Classification of Cellular Automata Using Asymmetry Graphs. GRIN v341659, 2016.

3. Goldberg, M. 'Complexity.' Telicom XY.19, September 2002, pp 44-48. Journal of the International Society for Philosophical Enquiry, ISSN: 1087-6456.

4. Goldberg, M. Complexity Arising from Entropy Acting on Asymmetric Substrates. Telicom XII.22 June/July 1998, pp 38-40. Journal of the International Society for Philosophical Enquiry, ISSN: 1087-6456.

5. THREE PAPERS **1.** Sundelacruz S, Levin M, Kaplan, D.L. The Role of Membrane Potential in the Regulation of Cell Proliferation and Differentiation Stem Cell Rev. 2009 Sep; 5(3):231-46. DOI: 10.1007/s12015-009-9080-2. Epub, 2009 Jun 27. **2.** Levin, M. et al. Endogenous Bioelectrical Networks Store Non-Genetic Patterning Information during Development— Cracking the Bioelectric Code. The Journal of Physiology 2014-43. **3.** The bioelectric code: An ancient computational medium for dynamic control of growth https://doi.org/10.1016/j.biosystems.2017.08.009.

6. Scale-Free Network; complex networks Wikipedia.

7. TWO PAPERS. **1.** Levin, Martyniuk, The Bioelectric Code: An ancient computational medium for dynamic control of growth and form. https://doi.org/10.1016/j.biosystems.2017.08.009. **2.** Levin, M. Cancer, Bioelectrical Signals and the Microbiome Connected. Tufts University, May 27, 2014.

8. Watanabe, A. et al. Fractal and Small-World Networks Formed by Self-Organized Critical Dynamics. J. Phys. Soc. Japan (2015). http://dx.doi.org/10.7566/JPSJ.84.114003.

9. Moore, D., Walker, S., and Levin, M. Cancer as a Disorder of Patterning Information: Computational and Biophysical Perspectives on the Cancer Problem. Copyright 2017 IOP Publishing Ltd. Convergent Science Physical Oncology, Volume 3, Number 4.

10. Lobikin, M., Chernet, B., Lobo, D., Levin, M. Resting Potential, Oncogene-induced Tumorigenesis, and Metastasis: the Bioelectric Basis of Cancer *in vivo*. Published 29 November 2012, 2012 IOP Publishing Ltd. Physical Biology, Volume 9, Number 6.

11. Kayama, Y. Complex Networks Derived From Cellular Automata. Department Of Media And Information, BAIKA Women's University, 2-19-5, Shukuno-Sho, Ibaraki-City, Osaka-Pref., Japan

12. Seunghoon Oh, Bargiello, T. Department of Physiology, College of Medicine, Dankook University, Cheonan, Korea; Dominic P. Purpura Department of Neuroscience, Albert Einstein College of Medicine, Bronx, NY, USA. Voltage Regulation of Connexin Channel Conductance

13. Matthew D. B., Bassel, G.W. Network-based approaches to quantify multicellular development. Published 11 October 2017.DOI: 10.1098/rsif.2017.0484

14. Harris, A., Spray, A., and Bennett, M. Intercellular channels formed by connexins (gap junctions) are sensitive to the application of transjunctional voltage (V_j), to which they gate by the separate actions of their serially arranged hemi-channels (1981. J. Gen. Physiol. 77:95–117).

15. Lampe, P., Lau, A. Regulation of Gap Junctions by Phosphorylation of Connexins. Archives of Biochemistry and Biophysics 384 (2): 205-16, 2000.

16. Davies, J. Cellular Mechanisms of Morphogenesis. 2008 DOI:10.4249/scholarpedia.3615 Electric Gradients over a Field Guide the Geometry of an Organism.

17. Fuller, B., Cell Division 20105:5 https://doi.org/10.1186/1747-1028-5-5. BioMed Central Ltd., 2010.

18. Chipman, K., et al. 'Disruption of Gap Junctions in Toxicity and Carcinogenicity'. Toxicological Sciences, Volume 71, Issue 2, 1 February 2003, Pages 146-153, https://doi.org/10.1093/toxsci/71.2.146- how gap junctions can protect from spread of toxins.

19. Leithe, E., et al. Downregulation of Gap Junctions in Cancer Cells. Institute for Cancer Research. Review article. Loss of Gap Junction fosters cancer and cancer spread.

20. Brax, N., Amblard, F. 'A Self-Repairing Solution for the Resilience of Networks to Attacks and Failures'. https://www.researchgate.net/publication/230726048_A_self-repairing_solution_for_the_resilience_of_networks_to_attacks_and_failures. Mar 03 2018.

21. Goldberg, M. 'Complexity and Randomness,' Telicom XVI.3- June/July 2003 pp 65-67. Journal of the International Society for Philosophical Enquiry, ISSN: 1087-6456.

22. Tononi, G., From the Phenomenology to the Mechanisms of Consciousness: Integrated Information Theory 3.0 Published: May 08, 2014 DOI: 10.1371/journal.pcbi.1003588 Featured in PLOS Collections.

23. Stam, C., Jaap Reijneveld C. Graph theoretical Analysis of Complex Networks in the Brain. Nonlinear Biomedical Physics 2007 1:3 DOI: 10.1186/1753-4631-1-3.

24. Bullmore, E., Sporns, O. Complex Brain Networks: Graph Theoretical Analysis of Structural and Functional Systems. Nature Reviews. Neuroscience **10**, 186-198 (March 2009). doi: 10.1038/nrn2575.

25. Basalyga, G., Gleiser, P., Wennekers, T. 'Emergence of Small-World Structure in Networks of Spiking Neurons through STDP Plasticity' Volume 718 of the series <u>Advances in Experimental Medicine and Biology</u> pp 33-39

26. Neural Development Wikipedia.

27. Bassett, D. et al. Adaptive Reconfiguration of Fractal Small World Human Functional Networks PNAS: Vol103 no. 51 19518-19523 doi: 10.1073/pnas.0606005103.

28. Gilson, M. et al. Emergence of network structure due to spike-timing-dependent plasticity in recurrent neuronal networks Biol. Cybern (2009)101:427–444 © Springer-Verlag 2009 Doi 10.1007/s00422-009-0346-1

29. Butz, M, et al. Homeostatic structural plasticity increases the efficiency of small-world networks. Front. Synaptic Neurosci 01 April 2014. https://doi.org/10.3389/fnsyn.2014.00007.
30. Cohen, R., Havlin, S. Scale-Free Networks Are Ultrasmall. Physical Review Letters Vol. 90 Number 5 (week ending Feb. 2003). Minerva Center And Department Of Physics, Bar-Ilan University, Ramat-Gan, Israel.

31. Complex Networks Wikipedia.

32. Andresen, C. A. Properties of Fracture Networks and Other Network Systems 2008 Thesis for PhD Norwegian University of Science and Technology, Dept. of Physics.

33. Bullmore, E., Sporns, O. The Economy of brain network organization-Nature Reviews. Neuroscience **13**, 336-349 (May 2012). Doi10.1038/nrn3214

34. Spach, J. et al. 'Mechanism of origin of conduction disturbances in aging human atrial bundles: Experimental and model study.' Gap junction dysregulation in aging leads to atrial tachyarrythmias.

35. Wright, J.A. et al., Connexins and Diabetes   Department of Vascular Surgery and Department of Cell and Developmental Biology, University College London Hospital, 74 Huntley Street, London NW1 2BU, UK

36. Sato, T., et al., Downregulation of Connexin 43 Expression by High Glucose Reduces Gap Junction Activity in Microvascular Endothelial Cells in Retina (produces diabetic retinopathy). Department of Ophthalmology Boston University School of Medicine, Boston, Mass   Impaired intercellular communication may contribute to breakdown of homeostatic balance in diabetic microangiopathy. Diabetes 51:1565–1571, 2002

37. Dorin, A. 'Physically Based, Self-Organizing Cellular Automata.' school of computer science and software engineering- Monash University, Clayton Australia 3168

38. Gavrilov, L., Gavrilova, N. The Reliability Theory of Aging and Longevity Journal of Theoretical Biology, 2001 System theory of aging

40. Barron, A. Annals of Probability. Entropy and the Central Limit Theorem, Volume 14, Number 1 (1986), 336-342.

41. 'tHooft, G. Entangled quantum states in a local deterministic theory 2009 arXiv:0908.3408

42. Yoshihiko Kayama. Network representation of cellular automata Article · April 2011 DOI: 10.1109/ALIFE.2011.5954643.

43. Oh, S., Bargiello, T., Purpura, D. 'Voltage Regulation of Connexin Channel Conductance' Department of Physiology, College of Medicine, Dankook University, Cheonan, Korea, and Department of Neuroscience, Albert Einstein College of Medicine, Bronx, NY.

44. Pietak A, Levin M. 2017 Bioelectric gene and reaction networks: computational modeling of genetic, biochemical and bioelectrical dynamics in pattern regulation. J. R. Soc. Interface 14:20170425. http://dx.doi.org/10.1098/rsif.2017.

45. Jackson, M, et al. Network-based Approaches to Quantify Multicellular Development October 11, 2017 DOI: 10.1098/rsif.2017.0484

46. Akberdin, I.R., et al. A Cellular Automaton Model of Morphogenesis in Arabidopsis Thaliana Institute of Cytology and Genetics, Siberian Branch, Russian Academy of Sciences, Novosibirsk, Russia b Novosibirsk State University, Novosibirsk, Russia c Novosibirsk State Technical University, Novosibirsk, Russiae-mail: akberdin@bionet.nsc.ru https://www.researchgate.net/publication/226035965_A_cellular_automaton_model_of_morpho genesis_in_Arabidopsis_thaliana [accessed Mar 08 2018].

47. Mosteiro, L. Tissue damage and senescence provide critical signals for cellular reprogramming in vivo. Science 25 NOV 2016 Vol 354, Issue 6316 DOI: 10.1126/science.aaf4445

48. Negentropy. Wikipedia

49. Spray, D. et al. Gap junctions and Bystander Effects: Good Samaritans and executioners. Online 2012 December 11 DOI: 10.1002/wmts.72 Wiley Interdiscip Rev Membr Transp Signal. 2013 January/February; 2(1): 1–15

50. (January 1996) Gap junctions play a role in the 'bystander effect' of the herpes simplex virus thymidine kinase/ganciclovir system in vitro. Gene Ther. 3 (1): 85-92.

51. Forever young? Science 2, February 2018 pp 506-7 Naked Mole Rats may know the secret.

52. Green, C. et al. Interrupting the inflammatory cycle in chronic diseases – Do gap junctions provide the answer? https://doi.org/10.1016/j.cellbi.2008.09.006

53. Gap junctions. Wikipedia

54. Carroll, B. et al., Persistent mTORC1 signaling in cell senescence results from defects in amino acid and growth factor sensing , UK55. Hilary L. Ashe, James Briscoe. The interpretation of morphogen gradients
Development 2006 133: 385-394; doi: 10.1242/dev.02238

56. Chipman, Kevin J. et al. Disruption of Gap Junctions in Toxicity and Carcinogenicity
Toxicological Sciences, Volume 71, Issue 2, 1 February 2003, Pages 146–153,
https://doi.org/10.1093/toxsci/71.2.146
57. Spray, David, et al. Gap junctions and Bystander Effects: Good Samaritans and executioners
Wiley DOI: 10.1002/ December 11. DOI: 10.1002/wmts.72

58. Levin M[1], Stevenson CG. Regulation of cell behavior and tissue patterning by bioelectrical signals: challenges and opportunities for biomedical engineering. Annual Review Biomedical Engineering 2012.14:295-323-doi: 10.1146/annurev-bioeng-071811-150114.

59. Wei Zhao Zhong. Cisplatin-induced premature senescence with concomitant reduction of gap junctions in human fibroblasts-March 2004 DOI: 10.1038/sj.cr.7290203 Peking University

60. TWO REFERENCES: **1.** John Wheeler. **2.** Digital Physics Wikipedia

61. THREE REFERENCES: **1.**Gerard 'tHooft. The Cellular Automaton Interpretation of Quantum Mechanics https://arxiv.org/pdf/1405-1548 pdf **2.** Superdeterminism Wikipedia. **3.** Counterfactual definiteness Wikipedia.

62. Science 16 February, 2018 pp 738-9 Is Evolution Predictable?

63. Takahashi, K; Yamanaka, S (2006). "Induction of pluripotent stem cells from mouse embryonic and adult fibroblast cultures by defined factors". Cell 126 (4): 663-76. Doi: 10.1016/j.cell.2006.07.024 PMID 16904174.

64. Steenbuck, D. et al. Self-organizing small-world networks are most robust against local disturbances. Frontiers in Computational Neuroscience 2011

65. Self-organization. Wikipedia

66. Isaeva, V. Self-Organization in Biological Systems April 2012   DOI: 10.1134/S1062359012020069 · Source: PubMed

67. Autocatalysis. Autocatalytic set. Wikipedia

68. Wong, R. et al. Role of Gap Junctions in Embryonic and Somatic Stem Cells. In Stem cell reviews 4(4):283-92 · September 2008 DOI: 10.1007/s12015-008-9038-9 · Source: PubMed

69. Holder, J.W. et al. Gap Junction Function and Cancer American Association for Cancer Research DOI: August 1993

70. Trosko, J.E., Ruch, R.J. Gap Junctions as Therapeutic Targets https://msu.edu/~trosko/Manuscripts/Ruch_gap_junctions/Ruch final

71. Levin, S. Science Daily Bioelectrical signals turn stem cells' progeny cancerous; newly discovered 'instructor cells' can deliver deadly directions. Tufts University: October 19, 2010

72. Levin, M., Zhang, Y. Particle tracking model of electrophoretic morphogen movement reveals stochastic dynamics of embryonic gradient Center for Regenerative and Developmental Biology, The Forsyth Institute, and Department of Developmental Biology, Harvard School of Dental Medicine, Boston, Massachusetts

73. Lu, Feng; Gao, JianHua et al. Variations in Gap Junctional Intercellular Communication and Connexin Expression in Fibroblasts Derived from Keloid and Hypertrophic Scars Plastic and Reconstructive Surgery: March 2007 - Volume 119 - Issue 3 - p 844-851 DOI: 10.1097/01.prs.0000255539.99698.f4

74. THREE REFERENCES: 1.Anuj Kumar. An Overview of Nested Genes in Eukaryotic Genomes 19 June 2009, doi: 10.1128/EC.00143-09 2.Nested genes Wikipedia. 3.Raquel Assis, Alexey S. Kondrashov   Nested genes and increasing organizational complexity of metazoan genomes Published online 2008 Sept 5; doi: 10.1016/j.tig.2008.08.003

75. Ben-Zion et al. Creating gradients by morphogen shuttling https://doi.org/10.1016/j.tig.2013.01.001

76. FIVE ARTICLES: 1. McNally J.G., Scale-invariant pattern in the alga Micrasterias. Inst. for Biomedical Computing and Department of Biology Washington University St. Louis. 1990. 2. Papageorgiou, S., Venieratos, D. A reaction-diffusion theory of morphogenesis with inherent pattern invariance under scale variation. Journal of Theoretical Biology 983 Jan 7; 100(1):57-59. 3. Shalygo, Yuri. The Kinetic Basis of Morphogenesis. Gamma Ltd. Vyborg, Russia. <juri.shalygo@gmail.com>. 4. Yakimov, B.N., Scale invariance of Biosystems: From embryo to community. Russian Journal of Developmental Biology. May 2014, Volume 45 Issue 3, pp 168-176. 5. Umulis, David M. Analysis of dynamic morphogen scale invariance. Journal of the Royal Society Interface. March 11 2009.DOI: 10.1098/rsif.2009.0015.

77. Willebrords, J. et al. Structure, regulation and Function of Gap Junction in Liver. 22 March 2016 https://doi.org/10.3109/15419061.2016.1151875.

YOUR KNOWLEDGE HAS VALUE

- We will publish your bachelor's and master's thesis, essays and papers

- Your own eBook and book - sold worldwide in all relevant shops

- Earn money with each sale

Upload your text at www.GRIN.com and publish for free